The Key to Your Weather Forecast

A Field Guide

John E. Klein

PublishAmerica
Baltimore

First printing

ISBN: 1-4137-6592-0
PUBLISHED BY PUBLISHAMERICA, LLLP
www.publishamerica.com
Baltimore

Printed in the United States of America

Table of Contents

Introduction

Welcome to *The Key To Your Weather Forecast.* This field guide has been designed for quick and easy weather forecasting without the need for prior instruction or aid. Whether you use this guide in a remote outdoors location or just around town, this practical key makes it possible for anyone to forecast the weather in just a few minutes.

Today, talk of the current weather forecast is one of the most common but least understood topics. In a short time you will acquire an understanding of what causes the dramatic weather changes that we all talk about. In addition, you will be able to make your own exciting predictions that will be more accurate and up to date than the latest regional forecast from your local news station.

This book has been designed to make weather forecasting fun and easy. Most of the unnecessary technical language that is not important in learning how to accurately forecast the weather has been eliminated. There are many books that attempt to teach about the weather. The student is required to plow through page after page of text and diagrams, finally quitting out of sheer boredom or because of the seeming magnitude and complexity of the study of weather. The key makes weather forecasting simple and distinguishes this book from every other on the subject. With a minimum of effort, you can forecast the weather by answering the simple questions in the key. The promise of this book is to enable you to accurately forecast the weather of the next few days in about five minutes.

This key has not been designed to take into account every possible variation nature can throw at us or conditions that rarely occur. The primary focus has been to incorporate dominate regional forces that make forecasting so exciting and predictable. By using the key you will discover how easy it is to acquire an understanding of the major weather phenomena.

This book and key are specifically designed with the western United States in mind. For general forecasting, it is applicable to the balance of North

America and for that matter, the Northern Hemisphere. Conditions that are confined or unique to an area outside the western United States are not included, specifically the hurricane season in the southeastern United States and the subtropical weather that occurs in Florida. This key is not designed to forecast weather in the tropics or sub-tropics.

You are now ready to take the first step enabling you to forecast the weather. Because this book should not be read, but rather, used, it is important to preview the following chapter: "How to Use This Book."

How to Use This Book

The Key to Your Weather Forecast is divided into two sections. By using them properly, you will find that forecasting is both simple and exciting.

The key has been designed so you can quickly and easily make a weather forecast. It works in a follow-the-bouncing-ball format, leading you by a series of easy-to-answer questions. Usually after answering about four questions, you will be given the weather outlook.

By using the key regularly, you will become aware of changes going on in the sky. This will give greater insights such as alerting you to a storm that takes a sudden course change, causing your location to miss most or all of the precipitation. Many times I have made more up to date and correct adjustments to the news media's weather forecasts while monitoring conditions and using the key.

If you desire to know why the weather acts the way it does and the principles that lie behind the key, the chapter "Let's Talk About the Weather" is for you. The study of weather can be fun. This chapter will show you that it is easy, too!

The weather glossary will help you understand the meaning of new words and phrases. I have attempted to eliminate all but the essential vocabulary needed to understand the weather.

The Key

You are about to make your first weather forecast! To be sure it is as accurate as possible, it will be important to use the key properly. Let's take a minute to get acquainted with it.

The key operates by guiding you through a series of questions, usually not more than four. Starting with the first question, you will be given several answers from which to choose. Yes, it's multiple choice all the way. No fill in the blanks or essay questions here! Once you make a multiple choice selection, you will be directed to another section of the key and another easy question.

The sections are named after the type of question found there: season, wind, clouds, barometric pressure, storm front, and outlook. The name of the section to which you are directed will be followed by a number, which leads to a specific question in that section. For example, if you were directed to Wind 3, that would mean you should turn to the wind section and answer the third question.

Continue answering questions as directed until reaching the outlook. If your outlook is forecasting clear weather, check the conditions once a day for possible changes. If unsettled conditions have been forecast, you may need to monitor conditions several times a day. The outlook will tell you when to use the key again.

The first question involves the seasons of the year. Most people are well aware of what season it is. Didn't I tell you this was going to be easy?

OK, let's get started! To begin, answer the question *Season 1* below.

Start here.

Season

Season 1. What season of the year is it?

Your four choices are:

a. Winter (Dec. 21 – March 20)* ......See Wind 1
b. Spring (March 21 – June 20)* ......See Wind 2
c. Summer (June 21 – Sept. 20)* ......See Wind 3
d. Fall (Sept. 21 – Dec. 20)*......See Wind 4

* Actual dates vary a day or two from year to year.
Go *now* to the wind question to which you were directed above.

Season 2. You have observed south or southeast winds. Answer the following question. Is the current month March, April, or May?

a. Yes ......See Clouds 3
b. No ......See Clouds 6

Wind

When you read a wind direction such as west, always understand that to mean only west. Westerly, however, would include southwest, west, or northwest. Adding the suffix "-erly" implies a broader meaning. All movements of the air will be called wind. This will include slight breezes to gale force winds. Sometimes the air along the ground will come from a different direction than the air (clouds) in the low level of the atmosphere. In this case, always use the wind direction of the low-level atmosphere. See *the atmosphere and clouds* sections in "Let's Talk About the Weather."

Wind 1. It is winter or possibly spring or fall. Answer the following question. What direction is the surface wind coming from?

a. North or northwesterly ......See Clouds 1
b. East or northeast ......See Clouds 2
c. Southerly ......See Clouds 3
d. No wind ......See Outlook 5

Wind 2. It is spring (March 21 – June 20). Answer the following question. What direction is the surface wind coming from?

a. Northwesterly ......See Clouds 1

b. Northeast ......See Clouds 2
c. Southwest ......See Clouds 3
d. South, southeast or east ......See Season 2
e. No wind ......See Outlook 5

Wind 3. It is summer (June 21 – September 20). Answer the following question. What direction is the surface wind coming from?

a. Northerly ......See Storm Front 1
b. West or southwest ......See Storm Front 2
c. South, southeast, or east ......See Outlook 8
If you are in the eastern U.S. ......See Outlook 3
If you are in the north central U.S.......See Clouds 3
d. No wind ......See Outlook 5

Wind 4. It is fall (September 21 – December 20). Answer the following question. What direction is the surface wind coming from?

a. Northwesterly ......See Clouds 1
b. East or northeast ......See Clouds 2
c. Southerly ......See Clouds 3
d. No wind ......See Outlook 5

Wind 5. Clear skies and no winds or northwesterly to east winds are observed. Answer the following question. What conditions prevail?

a. Strong, hot northeasterly winds, and you are in the southwest United States ......See Outlook 9
b. Northwesterly winds ......See Storm Front 1
c. Northeast or east winds and you are not in the southwest United States......See Outlook 6
d. No wind ......See Outlook 5

Clouds

If there is more than one type of cloud present, pick the most abundant type, or better yet, pick the type that has appeared most recently from the horizon the clouds are coming from. Ignore new cumulus clouds that form around mountains. When lenticular and other cloud types (see glossary) are present, always read both of the outlook options. The outlook given by following the lenticular question will always be less important. Refer to the cloud pictures at the end of this book to make positive identification of the cloud types.

Clouds 1. You have identified the wind direction as northwesterly. Answer the following question. What types of clouds are present? You may choose any answer and/or "h" if applicable.

a. Cumulus ......See Barometric Pressure 1
b. Stratus ......See Barometric Pressure 1
c. Altostratus ,.....See Outlook 3
d. Cirrostratus ......See Outlook 3
e. Cirrus ......See Outlook 4
f. Stratocumulus ......See Storm Front 3
g. Cirrocumulus ......See Outlook 3
h. Lenticular ......See Clouds 4
i. Mostly clear ......See Wind 5

Clouds 2. You have identified the wind direction as east or northeast. Answer the following question. What types of clouds are present? You may choose any answer, and/or "e" if applicable.

a. Cumulus or stratus ......See Barometric Pressure 1
b. Stratocumulus or altostratus ......See Barometric Pressure 2
c. Cirrus, cirrocumulus, or cirrostratus ......See Barometric Pressure 2
d. Clear or mostly so ......See Barometric Pressure 4
e. Lenticular ......See Clouds 4

Clouds 3. You have identified the wind direction as south, southeast, or southwest. Answer the following question. What types of clouds are present? You may choose any answer and/or "e" if applicable.

a. Cumulus or stratus ......See Outlook 3
b. Cirrostratus or cirrocumulus ......See Outlook 3
c. Altostratus or altocumulus ......See Outlook 3

d. Cirrus ......See Outlook 4
e. Lenticular ......See Clouds 5
f. Clear or mostly so ......See Barometric Pressure 3

Clouds 4. If lenticular clouds and winds from the north or northwest have been observed, see Outlook 1. If lenticular clouds and winds from the east or northeast have been observed, answer the following question. Are the lenticular clouds getting bigger or smaller?

a. Smaller ......See Outlook 1
b. Bigger ......See Outlook 2

Clouds 5. You have observed lenticular clouds and southerly winds. Answer the following question. Are the lenticular clouds getting bigger or smaller?
a. Smaller ......See Outlook 7
b. Bigger ......See Outlook 3

Clouds 6. You have observed south, southeast, or east winds and the month is June. Answer the following question. Are there clouds of any type approaching from a westerly direction?

a. Yes ......See Clouds 3
b. No ......See Outlook 8

Barometric Pressure

If you do not have access to a barometer, you will have to watch other signs to answer the barometric pressure questions. Watch for increasing clouds, persistent southerly or easterly winds, or winds becoming more southerly or easterly. In these cases, if you do not have access to the pressure readings, the answer to all barometric pressure questions will be decreasing. Otherwise, assume that the pressure is rising and the answer to all barometric pressure questions then will be increasing.

Barometric Pressure 1. You have observed any type of cumulus or stratus clouds and a northwesterly or a northeasterly wind. Read your barometer and answer the following question. Is the pressure increasing, decreasing, or flat?

a. Increasing ......See Outlook 1
b. Decreasing or flat ......See Outlook 2

Barometric Pressure 2. You have observed cirrus or stratocumulus or stratus clouds at any height, or cumulus at any height, or clear skies with east or northeast winds. Read you barometer and answer the following question. Is the pressure increasing, decreasing, or flat?

a. Increasing ......See Outlook 1
b. Decreasing or flat ......See Outlook 2

Barometric Pressure 3. You have observed southerly winds with a clear sky. Read you barometer and answer the following question. Is the pressure increasing, decreasing, or flat?

a. Increasing or flat ......See Outlook 7
b. Decreasing ......See Outlook 6

Barometric Pressure 4. You have observed east or northeast winds and clear skies. Read you barometer and answer the following question. Is the pressure increasing, decreasing or flat?

a. Increasing or flat but reading above 30.00" ......See Wind 5
b. Decreasing or flat but reading below 30.00" ......See Outlook 11

Storm Front

Storm Front 1. You have observed northerly winds. Answer the following question. Has a low pressure storm system passed by your area in the last few days? If so, recent conditions may have included precipitation, falling barometric pressure, southerly winds, and the normal forewarning clouds such as higher level stratus or higher level cumulus clouds (see "The Low-Pressure Storm System").

a. Yes ......See Outlook 1

b. No; if you are in the southwest United States, it is late summer or fall, and conditions are very windy ......See Outlook 9

c. No; all other conditions ......See Outlook 10

Storm Front 2. You have observed west or southwest winds. Answer the following question. Are there any of the signs typical of a low-pressure storm system approaching: falling barometric, forewarning cloud types (see "The Low Pressure Storm System").

a. Yes ......See Outlook 3
b. No; if you are in the north central United States and it is summer......See Outlook 8
c. No; if you are in the eastern United States and the winds are southwest......See Outlook 3
d. No ......See Outlook 10

Storm Front 3. Stratocumulus clouds and wind from the north or northwest have been observed. See Outlook 2. Don't expect much from this storm unless other storm precursor clouds are observed within hours (see "The Low-Pressure Storm System") and the barometric pressure starts to drop soon. If so, see Outlook 3. Otherwise, expect clearing and cooler conditions.

Outlook

All references to rain include snow if the temperature is below thirty-two degrees. References to precipitation can include any type. Watch the temperature to determine what type of precipitation may fall. Each outlook will direct you when to use the key again. This will usually be the most likely time to observe changes in the forecast. If you notice changes before the next suggested time, use the key immediately.

Outlook 1. Clearing and Cooler

The Forecast
Expect clearing skies, cooler temperatures, and possibly increased wind speeds. If precipitation is present, watch for it to taper off and stop over the next twenty-four hours. (In the mountains, this tapering off period will take longer.) Expect cooler temperatures over the next two days. In fog-prone areas, these more stable conditions may allow fog to develop.

Taking a Closer Look

More than likely a storm system (with or without rain accompanying it) has passed through your area in the last day. If precipitation is still falling, watch for it to taper off as the storm leaves. If you have easterly winds, the storm center now lies south or southeast of your location. If the storm is strong enough, it is possible a few residual showers may come soon. Otherwise, expect clearing skies, cooler temperatures, and possibly increased wind speeds.

If no storm has passed recently, clear or clearing skies can still be expected for at least a few days before the next possible storm will appear, unless changing conditions dictate otherwise. The next most likely time for the arrival of a storm is about three to four days. To determine the exact day for the storm to arrive start counting days from the arrival of your last storm system (the day the precipitation started) or from the observance of the very first signs of the approach of a new storm.

In valleys that lie to the east of the mountain chains along the West Coast, cold air can get trapped along the ground. This trapped air can cause the unusual condition of light precipitation falling, with slight northerly to easterly breezes and a rising pressure reading. Usually fog or low clouds will form overhead and humidity readings will be above sixty percent. These clouds will look like fog. They may lie along the ground or be overhead only. When sunlight starts to reach the ground, the fog will dissipate. Nightfall will usually reestablish these clouds as long as stable conditions, colder temperatures, and only light breezes prevail. This foggy condition will only occur from mid-fall to mid-spring.

If it is summertime, possibly a dry front or a front too far to the north has passed. Expect windier conditions, clearing skies, and then usually a return to the typical southerly to westerly breezes and warmer temperatures. Next time to use the key: tomorrow.

Outlook 2. Partly Cloudy, Maybe Rain

The Forecast

Expect cloudy skies and possible precipitation. This unsettled condition will last until the barometric pressure stops falling and goes back up. For temperature changes, see below.

Taking a Closer Look

A storm has just passed and now lies to the east of your location causing the northwesterly winds; or a storm has just passed to the south of your location causing the easterly winds. If the storm stalls and intensifies after it passes, it will cause the barometric pressure to fall. If a storm forms to the south or east of your location or approaches from the south, this too can cause northwesterly or easterly winds and falling barometric pressure. In these cases expect unsettled conditions until the barometric pressure begins to rise. If the barometric pressure continues to fall, increased cloudiness and precipitation are possible at any time. Cooler temperatures are the norm under these conditions. In the case of easterly winds, also see Outlook 11.

It is also possible that another storm system to the west or south of your location is quickly approaching. In this case, watch for the winds to shift around and become southerly or easterly respectively and the temperature to rise. Look for middle and upper-layer clouds to appear from the westerly horizon. Next time to us the key: six hours.

Outlook 3. Cloudy and Rainy

A storm is approaching. Which one of following statements is most true? Don't include clouds around mountains.

You have observed clouds mostly in the high layer.
See 1 below: Rain or Snow in Two to Three Days

You have observed clouds mostly in the middle layer of the sky. Very few, if any, low-level clouds are present.
See 2 below: Rain or Snow in One or Two Days

For all other observations, answer the following question.

Which condition is truer?
The winds are east to northeast. ......See 3 below: Unsettled, Maybe Rain or Snow
The winds are not east to northeast. ......See 4 below: Cloudy and Rainy

1. Rain or Snow in Two to Three Days

The Forecast

A storm front is approaching. Expect increasing clouds of all types, windier conditions, and warming temperatures over the next few days. The storm's arrival is about three days away. We will have to wait before forecasting precipitation.

Taking a Closer Look

With cirrus, cirrostratus, or cirrocumulus cloud types present, expect possible precipitation in about two to four days if the storm is strong enough and if the other signs of an approaching storm appear (see "The Low-Pressure Storm System"). If southerly winds do not persist, or northerly winds do not become southerly, the barometric pressure does not start falling, and other cloud types are fragmented or poorly formed, expect a weak storm. For additional information see Outlook 4. Next time to use the key: make another observation tomorrow and start in your key at Outlook 3 if the wind continues to be southerly. If the wind tomorrow is not southerly, start at the beginning of the key.

2. Rain or Snow in One to Two Days

The Forecast

A storm front is approaching. Expect increasing clouds, windier conditions, and warming temperatures. The storms arrival is about two days away.

Taking a Closer Look

With altostratus or altocumulus clouds expect precipitation in about one or two days if the storm is strong enough and if the other signs of an approaching storm appear (see "The Low-Pressure Storm System"). If southerly winds do not persist, the pressure readings do not start falling, and other cloud types do not appear, don't expect much from this weak storm. Next time to use the key: make an observation in about six hours. Start in your key at Outlook 3 if the wind continues to be southerly. If the winds are not southerly, start at the beginning of the key.

3. Unsettled, Maybe Rain or Snow

The Forecast

Expect cloudy skies and maybe precipitation over the next twenty-four hours. The days that follow will be predictable by reading "Taking a Closer Look" below.

Taking a Closer Look

If easterly winds are present, it is possible a storm system is passing to the south of your location or, if you live in the eastern United States, a storm system is approaching your location from the south. Watch the direction of the pressure. The larger the drop in pressure, the closer the storm front will come. If the storm comes close, expect the wind direction to shift more to the east. This increases your chance for significant amounts of precipitation over the next few days. If the pressure does not decrease much, the storm front will pass to the south of your location and not affect you as greatly. Northerly winds, cooler temperatures, and clearing skies would then be expected over the next few days. Review Diagram 1. Also see Outlook 6. Next time to use the key: six hours.

4. Cloudy and Rainy or Snowy

Which conditions have you experienced over the last few days?

You are in the eastern United States or Midwest; it is spring, summer or fall, and it is hot and muggy. ......See Outlook 12.

Conditions have been dry but with increasing signs of the approach of a storm front (see "The Low-Pressure Storm System"). ......See 4A below: Rain or Snow on its Way

A storm front has passed in the last two days. Possibly intermittent precipitation and low-level clouds have been the norm. ......See 4B below: Continuing Rain

4 A. Rain or Snow on Its Way

The Forecast

There is a good chance of imminent precipitation. Unsettled conditions will continue as long as the wind is southerly. Once the winds shift to northerly, the next few days will bring dropping temperatures, increasing wind speeds, and clearing skies.

Taking a Closer Look

The cumulus or stratus clouds should be covering more and more of the sky if they are to bring significant amounts of precipitation. The barometric pressure should be falling and reading less that 30.00". If so, expect precipitation within hours. If the barometric pressure is greater than 30.00", the storm is weak; do not expect a lot in the way of precipitation. Conditions should clear and the wind should shift toward the north soon. After the passage on a front, it may take a few hours before the winds shift from southerly to northerly. Watch closely, however. Persistent southerly winds after the passage of the front may indicate more storminess is on its way. Next time to use the key: six hours.

4 B. Continuing Rain or Snow

The Forecast

There is a good chance of imminent precipitation. Unsettled conditions will continue as long as the wind is southerly. Once the winds shift to northerly, expect temperatures to drop, wind speeds to increase, and skies to clear over the next few days.

Taking a Closer Look

Sometimes storms will be followed very closely by other storms. In these cases you might observe the following conditions:

* intermittent precipitation
* only partial clearing or no clearing at all
* the pressure falling, then slightly rising and then possibly falling again
* a persistent southerly wind.

The most important sign is the wind direction. If you observe the passing of a storm front but then do not see the typical changes (see "The Low-Pressure Storm System"), especially the shift from southerly to northerly winds, be forewarned, more unsettled weather can be expected. Precipitation could start at any time.

At this point there are three conditions that could account for unstable weather. First, a cold front just passed and another storm front is very closely following. Second, after the storm front approached, it stalled and is now stationary causing unsettled conditions until it moves out of your area. Third, the front that just passed is the warm front of a low-pressure storm system. Soon to follow is the cold front (see Diagram 1 in "Let's Talk About the Weather"). The third condition would have caused the following changes over the last few days:

* lowering pressure changing to flat or slightly rising pressure readings
* south to southeasterly winds changing to south to southwesterly winds
* cloud types changing from lowering stratus to both stratocumulus and cumulus in most likely a partially cloudy sky.

Expect these clouds to thicken with the approach of the cold front. Also, the temperature would have changed from cool to warmer readings with the passage of the warm front. Next time to use the key: six hours.

Outlook 4. A Storm Front Approaches

The Forecast

Look for warming temperatures, increasing winds, and more cloudiness over the next few days. There is a chance of precipitation in about three days.

Taking a Closer Look

You can expect a storm to arrive in about two to four days. You don't yet know if the storm will pass close enough or if it is strong enough to cause precipitation. A clue to the strength of the approaching storm can be learned by observing the direction of the jet stream. A south or southwest jet stream will increase the strength of the storm. North or northwest jet streams will weaken the storm. When westerly jet streams are observed, you must know the direction from which the jet stream was most recently coming. If the jet stream was coming out of a northerly direction before it shifted to the west, expect the storm to strengthen. If the jet stream was southerly before it shifted to the west, expect the storm to weaken. Look for warming temperatures and increasing winds over the next few days if the surface wind is or becomes southerly. Also see Outlook 3. Next time to use the key: tomorrow.

Outlook 5. Continuing Steady and Stable

The Forecast

Expect the next few days to be about the same as present conditions.

Taking a Closer Look

Conditions are stable. Watch for changes in the wind and sky. Those clues will warn you of changing conditions. Don't let still or negligible wind conditions catch you off guard. Sometimes these conditions occur just before a change in the weather. Conditions can change quickly. If it is April through October, and conditions are hot and muggy, also see Outlook 12. Next time to use the key: twelve hours.

Outlook 6. Here Comes a Storm from the South

The Forecast

Over the next few days expect an increase in wind speed, an increase in cloudiness, and an increasing chance of precipitation.

Taking a Closer Look

A storm probably is to the south of your location (this occurs more rarely in the southwestern United States) and is moving in a northerly or northeasterly direction. The more sharply the pressure is falling, the greater the chances of significant amounts of precipitation. Clearing conditions will be expected when the pressure begins to rise and the wind shifts to a more northerly direction. If stratocumulus or stratus clouds are observed, expect precipitation at any time. Higher level stratus or cirrus clouds indicate that the storm front, bringing a chance of precipitation, will take longer to arrive—about one to three days. If lenticulars are present, they will be growing in size if rain is on its way. If lenticulars are getting smaller, the possibility of precipitation is declining.

Clear skies and easterly winds may indicate a new storm is forming to the south or southwest of your location. Watch for changes closely. Unsettled conditions can develop into storms quickly. If your location is the Midwest or southeast United States, and it is April to October, clear skies, southerly winds, and dropping barometric pressure may forewarn of thunderheads. In this case, also see Outlook 12. Next time to use the key: twelve hours.

Outlook 7. A Weak Storm Approaches

The Forecast

Expect partly cloudy, then clearing skies, winds out of a northerly direction, and cooler conditions to prevail over the next few days.

Taking a Closer Look

A storm is approaching from the west. Watch the lenticulars and the barometric pressure closely. More than likely the storm is not very strong or the storm is passing too far to the north for it to influence your weather outlook much. Check your readings from the last few days to see if you observed a northwesterly jet stream. This should help confirm that a weak storm is approaching. Next time to use the key: twelve hours.

Outlook 8. Monsoon Season, Rain and Thunder

The Forecast

Expect hot, muggier conditions with thunderstorms and rain possibly developing as the day progresses (western United States only; all other locations see Clouds 3).

Taking a Closer Look

The development of thunderstorm clouds will subside as the evening turns to night. Monsoon conditions will continue for the next few days or until the wind direction changes. Usually a shift in the wind direction to the west, northwest or north will end the monsoon effect, especially if you are between the western coast of the United States and about five hundred miles inland. However, it is common for these thunderstorms to come and go during the summertime as the wind changes.

If it is the last month of spring, remember the monsoon season usually arrives in early July and lasts for a few months (see Diagram 5). On occasion the monsoon season will arrive early, but this is not normally the case. The days preceding the arrival of the monsoon should be hot and muggy. Next time to use the key: tomorrow.

Outlook 9. Clear and Warmer, but Windy

The Forecast

Expect clear skies, warming temperatures, and winds coming out of a northerly direction over the next few days.

Taking a Closer Look

In the fall or late summer (more rarely in the winter) a condition will sometimes develop that causes moderate to strong northerly winds. This condition is common to southern California. It is accompanied with clear skies, hot dry air, and somewhat higher barometric pressure and is caused by high pressure that develops over the Oregon, Nevada, and Idaho region. The winds will die down as the high pressure to the north decreases or moves. In the northwest region, this high pressure produces clear skies and stable dry conditions. Next time to use the key: two days.

Outlook 10. Clear and Stable Summer Days.

The Forecast

Expect clear and dry conditions over the next few days.

Taking a Closer Look

This is a normal summertime condition. Expect clear and stable conditions. If you live along the West Coast, the breezes will bring cooling relief from the summer heat. If you live in the interior, always be on the lookout for southeasterly to southwesterly winds that bring thunderstorm activity. Next time to use the key: two days.

Outlook 11. Instability Is Coming

The Forecast

Conditions are unstable. There is an increasing chance for precipitation to come at any time (zero to three days).

Taking a Closer Look

An area of low pressure is probably forming near your location. Watch for any or all types of clouds to appear suddenly, especially cumulus or stratus. If the barometric pressure is also falling dramatically, significant cloud development and precipitation could be expected very soon (zero to two days). This condition requires close monitoring. Changes can occur suddenly.

It is also possible that a storm system is approaching. In this case, the arrival of low clouds and a chance for precipitation (one to three days) will occur after the arrival of some higher level forewarning clouds. Next time to use the key: six hours.

Outlook 12. Thunderstorms Are Coming

The Forecast

Conditions are unstable. With continuing southerly winds, watch for the development of large thunderheads (cumulus clouds) overhead or approaching usually from a southerly direction. Expect rain, hail, wind, thunder, and lightning. Tornados sometimes form from these large thunderheads. Next time to use the key: six hours.

Taking a Closer Look

The development of thunderstorm clouds will usually subside as the evening turns to night. These conditions will continue for the next few days or until the wind direction changes. Usually a shift in the wind direction to west, northwest, or north will end these storms. As these large thunderstorms move overhead, they will effect the wind locally. Do not confuse these swirling, changing wind directions caused by these thunderstorms. After they pass, make a new wind observation. If the wind direction resumes its southerly direction, assume more of these storms are on the way. If not, more stable conditions are coming. However, it is common for these thunderstorms to come and go during the summertime as the wind direction changes.

Let's Talk About the Weather

Ever since I can remember, I've enjoyed observing and trying to predict the weather. It was a curiosity of mine to discover various changes in the wind and sky as rain or snow storms came and went. This led to a lot of questions and a lot of time spent studying the answers to those questions. All this effort brought about a practical understanding of why the weather does what it does. I think you will be amazed after studying this section and becoming familiar with the key that weather forecasting can be so simple. Yes, predicting the weather far into the future is more difficult. But usually, all we really care about is what is going to happen today or tomorrow or over the weekend. That is exactly the scope of this book.

The Low-Pressure Storm System

In the United States during the fall, winter, and spring seasons (also summer in the eastern United States) the main cause of precipitation is the low-pressure storm system (See Diagram 1). An area of low pressure is always found at the center of a low-pressure storm system. In general, the lower the pressure, the stronger the storm system and the more precipitation it will bring. The closer the storm comes to your location, the greater the impact it will have on your weather and plans. Greater wind speeds can also be anticipated as the storm approaches. When confronted with an approaching storm system, what we need to know is: how close the storm will pass and how strong it is.

Storms may come from the north, south, or westerly directions. Rarely do they come from any easterly direction. (When you read a wind direction such as "west," always understand that to mean only west. Westerly, however, would include southwest, west, or northwest. Adding the suffix "-erly" implies a broader meaning.) However, if a storm is followed over a greater distance, it will be found to almost always proceed ultimately in a easterly direction. With that in mind, let's look at Diagram 1. Note the "L" which denotes the center of low pressure and the center of the storm system. The rings around the "L" indicate areas of different pressure, much like the lines on a contour map denote different altitudes. As you move farther from the center of the low, the pressure increases and the less of an impact the storm system will have in your area. The arrows in Diagram 1 show the direction of the surface wind. Wind will go in a counter-clockwise direction around the low. The arrows point slightly inward, because air will always attempt to go from an area of higher pressure to an area of lower pressure. This is similar to a tank filled with air under high pressure. If this tank is punctured, the gas inside will flow out of the tank.

The circular lines in Diagram 1 vary in how far apart they are from one another. Each circular line, as you proceed away from the low, can be viewed also as demarcations of equal amounts of energy affecting the air mass between the lines. This means that whether the area between each line is narrow or wide, the same amount of energy is affecting those areas. What affect does this have in areas where the pressure lines are far apart verses areas with lines closer together?

An example involving two balls will make this clear. The first was made of solid rubber and was one foot wide. The second ball was also made of solid rubber but was ten feet wide. If you were to push with equal force on both balls, which would move the fastest? Of course the small ball, having a smaller mass, would move fastest. In like manner, in the area where the lines were closer and the amount of air less, the air would move at a higher speed. It's windiest where the lines are closer together. Therefore, the windiest times will be just before and just after the passage of a front.

If you lived to the east of this approaching storm, (the storm is moving from left to right across the page), your location might be where the letters "C," "F," or "I" appear in relation to the low-pressure storm system. Which location has the lowest pressure? Location F. It lies directly in the path of the center of the low. The other two locations (C and I) are not yet seeing much in the way of reduced pressure.

Diagram 1

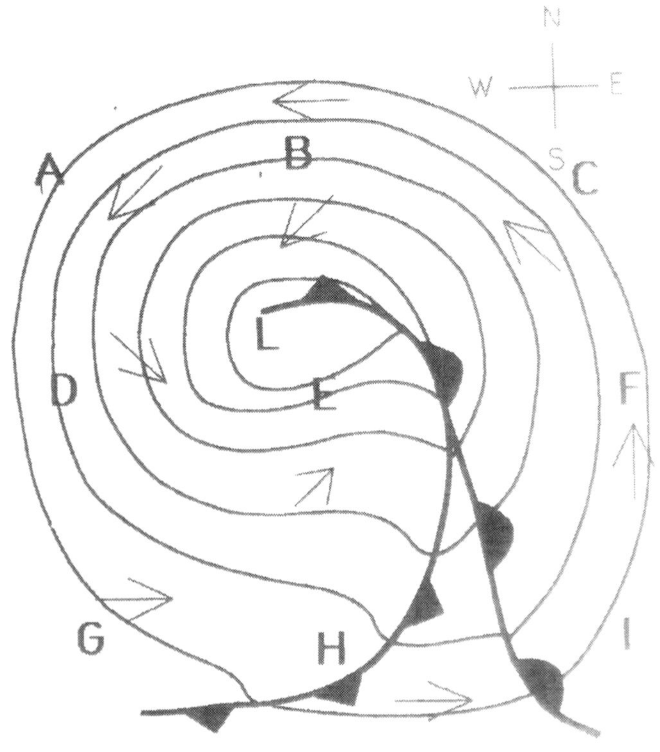

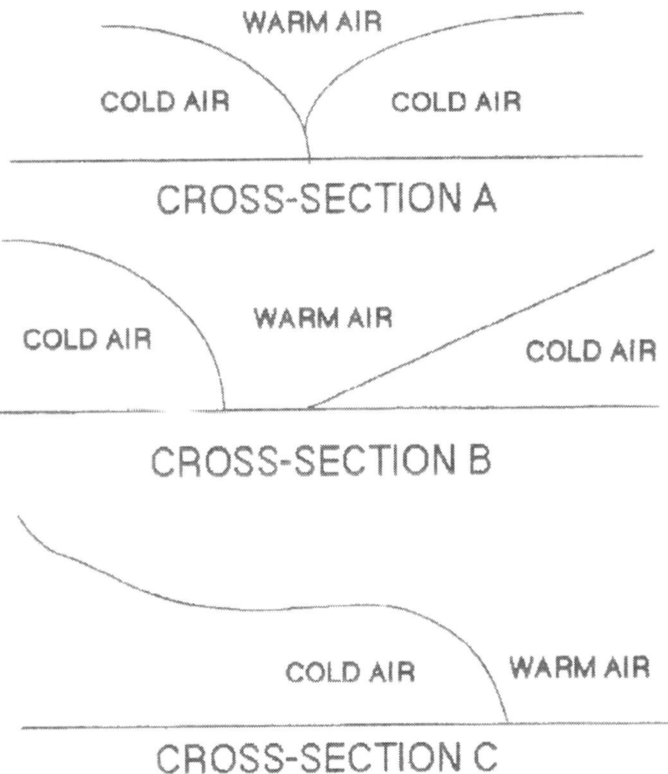

Which location has the most southerly winds? Again, it's Location F. Location I has southwesterly winds (as can be seen by the direction of the arrow closest to I), indicating that the center of the storm will pass to the north of the "I" location. The person that was at Location C would conclude that a storm is approaching, but the center may pass to the south because of the easterly winds.

Let's pretend we could, by snapping our fingers, move our location to the left or west on Diagram 1. That would mean, if we were at Location C, F, or I we would now be where the letters "B," "E," or "H" are respectively. What has happened to the pressure? All three of the new locations to the left (or west) show lower readings on the barometers. We see the largest drop in barometric pressure when we moved from the "F" location to the "E" location. Note also the changes in wind directions at the three new locations as compared to the locations we were at originally.

Ok, let's snap our fingers one more time to move our location again to the west. We have now appeared at the most westerly location shown on Diagram 1. This would be Location A, D, or G, along the left side of the page. If we had started at Location C, we would have visited Location B, and now be at Location A. If we had started at Location F, we would have visited Location E, and now be at Location D. If we had started at Location I, we would have visited Location H, and now be at Location G.

Take note of the changes in the wind direction and pressure as we have moved to the left or west from these three possible starting points. If we moved to the left from Location C, the wind direction would have been initially east, then northeast, and finally north. The pressure would have gone down a little, and then gone back up. Starting at Location F and moving left or west, the wind direction would have been initially south, then southwest, and finally north. The pressure would have gone down drastically, then go back up. Starting at Location I, and moving left or west, the wind direction would have been initially southwest, then west, and finally northwest. The pressure would have gone down a little, and then gone back up.

In real life, snapping our fingers doesn't get us much. We definitely can't change our location by doing it. We can learn from the imaginary trip we just took, though. The changes we have observed by moving to the left from the three initial starting locations are the same changes we would observe at these locations as a low-pressure storm system passes.

In our study of Diagram 1, we have assumed that the low center moved directly to the east from its present location. That is not always the case. Storms can come from the northerly, westerly, and southerly direction. However, some generalizations can be made about the wind changes and the pressure changes (see chart below).

Signs of the close passage of a storm verses a distant storm.

	Rain	Wind	Pressure
Passage of Close Storm	More	Greater Southerly or Easterly	Down then up
Not a Close Storm Passage	Less	Less Westerly	Less change Down then up

The more drastic the change in the pressure readings and the more southerly the wind direction, the closer the center of the storm will come to your location, and the more precipitation is likely to fall. After a low-pressure storm systems passes, the pressure will rise and the wind direction will turn northerly. Pressure readings that do not change much or go up and wind directions that never become southerly, but rather change to a northerly or more westerly direction are all signs that an approaching storm will not pass very close and will not impact your weather as much.

Many people think that storms come from the south because the wind comes out of the south when the storm approaches. We have seen that this is not the case. Southerly winds in Diagram 1 indicate the approach of a storm from a westerly direction.

The closer the low-pressure storm system comes to your location, the greater the chance for precipitation. The signs that indicate storms that are close (southerly winds and strong reductions of barometric pressure) also indicate greater chances of precipitation.

The Atmosphere

Clouds are one of the most important and obvious signs of an approaching low-pressure storm system. In order to use the key successfully, it will be very important to learn how to distinguish between the various types of clouds. This may seem like a tedious chore, but it's really not.

Before we start learning cloud types, let's first take a look at the part of the atmosphere where clouds occur. For our purposes, let's divide this part of the atmosphere into three layers: low, middle (alto), and high (cirro).

In the United States the low level is from the Earth's surface to about 6,000 feet. The middle layer is from 6,000 feet to about 23,000 feet. The high layer starts at about 16,500 and goes as high as 45,000 feet. As you go from north to south, the three layers get thicker and thicker. For example, in the tropics the high layer reaches 60,000 feet. This accounts for the overlapping of the ranges listed above.

Clouds

There are actually only two basic types of clouds. Both types occur in all three layers, giving us six different forms of clouds to learn. Refer to the cloud pictures at the end of this book to help you get acquainted.

The first basic type of cloud is the *cumulus* cloud. This cloud is caused by air rising quickly and takes the shape of cotton balls. Cumulus clouds may occur in the high, middle, and low layers of the sky or the same cloud may occur in all three at the same time. They may be small or huge (thunderclouds), but they will all be the same general type of cloud. Low cumulus clouds are the ones that bring precipitation.

As you compare the various cloud sizes, you will be able to determine the layer that the cloud is in. Very small cumulus clouds, about the size of a pea or smaller, are found in the high layer and are called cirrocumulus clouds. Larger cumulus, about the size of a golf ball, are found in the middle layer of the sky and called altocumulus. Cumulus clouds are the larger, low lying cotton-ball shaped clouds. Each type may occur in sheets, but not always. Usually during the passage of cold fronts, low level cumulus clouds will form one large cloud mass, making it impossible to distinguish separate clouds. Only varying intensities of the downpours will signal the passing of individual low level cumulus clouds imbedded within the cloud mass.

Cumulus Clouds (Cotton Balls)

Type	Height	Appearance
Cirrocumulus	high	pea sized Usually in groups
Altocumulus	middle	golf ball sized Usually in groups
Cumulus	low	large Rain may occur

The second basic type of cloud is the *stratus* cloud. These clouds all form sheets and are caused by air rising slowly, as if flowing up an inclining plane (refer to Diagram 2). They are also caused by air flowing across a cooler surface such as the ground or ocean. All strata type clouds will be very uniform and do not form sharply-edged groupings of clouds within the sheet. Many times stratus clouds are opaque. These clouds also occur in all three layers of the sky. Sometimes rings will form around the sun when they are present. Fog is a typical example of stratus clouds. Cirrostratus (high layer clouds) can form huge sheets, covering the entire sky. Sometimes the only evidence of the presence of these clouds is a ring around the sun and/or a slight tinting of the sky. Be careful, these clouds are easy to miss. Altostratus (middle layer cloud) also take the form of a sheet, but they are much more noticeable. They will shroud the sun but usually not block it out completely. A light precipitation may fall from them. The stratus cloud will lie just above the ground. Usually they will block out the sun and often will produce steady precipitation.

Stratus (Sheets)

Type	Height	Appearance
Cirrostratus	high	ring around the sun Tinting of the sky
Altostratus	middle	shrouds the sun Possible light rain
Stratus	low	steady precipitation

Among the special clouds we will get to know are the *cirrus*, *stratocumulus*, and *lenticular* clouds. Cirrus clouds only occur in the upper layer of the sky. They take on a stringy appearance, looking like hair blowing in the wind. Stratocumulus clouds form layers or sheets but include clumps or rounded cloud fragments. They occur in the lower layer of the sky. Lenticular clouds are very unique. They are stationary, usually saucer-shaped, and have a tendency to occur over the tops of mountains or other clouds. These clouds are formed by air being caused to rise and then descend. The air cools and reaches the saturation point as it rises then warms back up as it descends and dissipates. Lenticulars will help predict the arrival and strength of approaching storm fronts. Lenticulars will usually appear before and after the passage of a storm system. These clouds will grow in size as the storm approaches and shrink as the storm passes. If the lenticular clouds start to shrink in size before the arrival of the storm, this usually means the approaching storm is not very strong.

Special Clouds

Type	Height	Appearance
Stratocumulus	low	sheets, clumps Rounded cloud fragments
Lenticular	middle to high	saucer-shaped layer cloud
Cirrus	high	hair blowing in the wind

Storm Front Forewarning Clouds

We have seen how a low-pressure storm system forewarns of its approach by changing the wind direction and the barometric pressure. The most obvious signs, however, are the clouds it sends out ahead of the actual arrival of the storm front. The chart on the next page will summarize this information nicely. The first clouds to be seen are usually the cirrus clouds. They come about three days before the storm front's arrival. Cirrostratus and cirrocumulus clouds come next. They arrive about two to three days ahead of the front. Altostratus and altocumulus appear about one day before the arrival of the storm front. Stratus, cumulus, and stratocumulus all can appear just before and during the passing of the front. Cumulus clouds will also develop after the passage of the front. The precipitation from these cumulus clouds will diminish with time.

Cloud Messages

Cloud Types	Messages
Cirrus	three days before the arrival of a storm front
Cirrostratus & cirrocumulus	two to three days before the arrival of a storm front
Altostratus & altocumulus	one day before the arrival of a storm front

Stratus, cumulus & stratocumulus	hours before and during the arrival of a storm front
Linticular	a few days before and after a storm front

To the informed, each type of cloud tells a story. The stories will warn of coming changes in the weather that may impact our plans.

In mountain locations expect a longer time of precipitation before clear conditions prevail. Sometimes more precipitation will be received in the mountains after the passage of the front due to residual instability. The cold air behind the cold front has been over the Pacific Ocean for days (West Coast only). It has had time to absorb a lot of moisture. When this air comes onshore, it is pushed over the mountain chains, sometimes violently by the increased wind speeds that come after the passage of a storm front. The air cools, clouds form, and usually rain or show is the result.

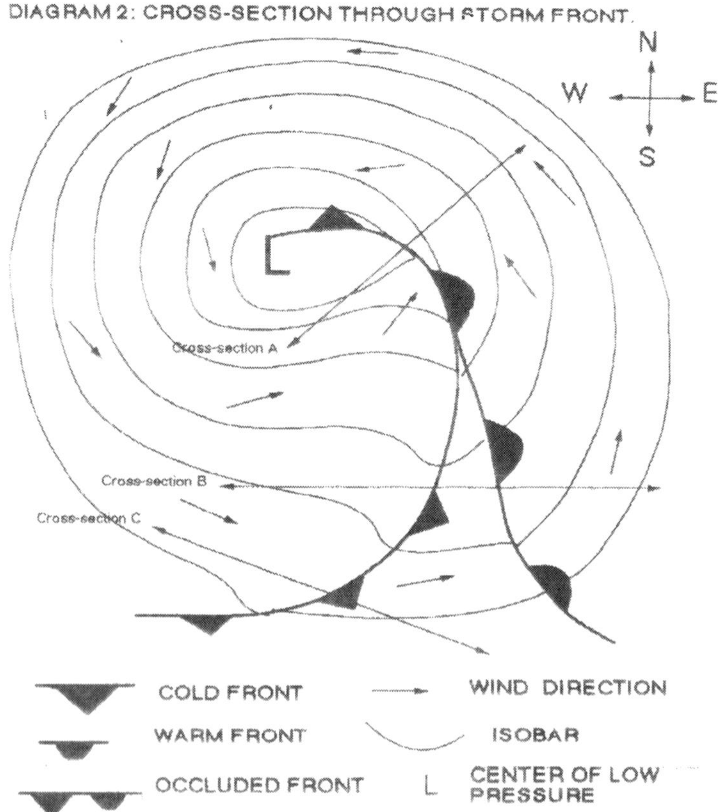

Warm Fronts, Cold Fronts, and Occluded Fronts

Refer to Diagram 1 and note two lines that look like curved spokes of a wheel. The line that is to the right (east) is a warm front and the line that is to the left (west) is a cold front. The warm front will always arrive first. The fronts are named after the type of air (cold or warm) that lies behind, or to the west or south of the front. All storm systems have these two fronts. Sometimes one or both fronts are not strong enough to be noticed or cause precipitation. This is especially true of the warm front in the southwestern United States. In order to observe the signs of the warm front, in that part of the country, usually the center of the low must pass closely by your location. This will be true even if the cold front is powerful. I have noticed that in the western United States, the warm front's passage occurs with various stratus mingled with mid-level cumulus clouds (altocumulus). The stratus will thicken but never cause much precipitation. Cumulus clouds will grow more dominate in the sky as the warm front passes and the cold front's influence dominates.

Warm fronts produce mostly stratus type clouds. Cold fronts produce cumulus-type clouds. This can be seen in Diagram 2, cross-section B. This line is a cross section of the sky through the two fronts. Notice the different cloud types and different temperatures on either side of the fronts. Since the winds are traveling from westerly to easterly in the area of the cross-section, the warm front is causing the warmer air to slide gradually up along a slowly eastwardly moving mass of colder air. These conditions cause stratus-type clouds to form. As the warm front approaches, stratus clouds will predominate and many times bring precipitation.

Cumulus-type clouds are formed by rapidly rising air caused by conditions present at the cold front. Notice the rapid ascension of the warm air to the east of the cold front. This is where cumulus clouds are formed and precipitation is most likely to occur and will be the heaviest. When both fronts are observed within a storm system, a period of clearing may occur between the departure of the warm front and the arrival of the cold front. Don't be fooled into thinking that the storm is over. Watch the wind, clouds, and pressure readings to warn of the cold front's approach.

Notice the front where both the warm and cold fronts have merged on Diagram 2. This is an occluded front. The cold front has overrun the warm front because cold fronts generally move at a greater speed than warm fronts. The warm air mass is not in contact with the ground any longer. Stratus, cumulus, and stratocumulus clouds are common along with the potential for stormy weather if this part of the front goes through the area. Weaker storms and storms that pass far to the north of your location usually include those storms that occur at the beginning and end of the rainy season. These storms will not send out all the signs we have talked about and the cloud types they do send out are not the best or largest examples. A strong storm front passing close to your location will almost always produce classic examples of forewarning cloud types and has a much greater chance of causing precipitation. These cloud formations are often large enough to cover the entire sky.

No matter what time of year it is, all approaching storm fronts, unless they are going to miss your location by a great distance, will forewarn of their arrival. The only exception is when a storm forms near your area. When this happens, cumulus clouds will usually form from a clear sky. But even under this condition, changes in wind direction and pressure will be observed. The change of the wind direction will depend on where the storm center is developing in relation to your location. Pressure readings will suddenly start to fall. Even though you did not get alerted by the normal sequence of cloud types, these sudden changes of pressure and wind direction and the immediate formation of cumulus clouds will alert you that something is making your weather unstable. The key will help you in evaluating these conditions and give the most likely outlook as the conditions change.

Dry Fronts

Most people will be unaware of the passage of a certain type of front called a dry front because no precipitation occurs. Not all storm fronts bring precipitation. Dry fronts are actually very weak fronts. The air masses on either side of the front did not differ in temperature much and/or the warm air mass did not contain much water vapor (the relative humidity was low). As you learn to identify the changing weather conditions around you, you will become aware of the passage of these dry fronts. Changes of wind speed, wind direction, temperature, and the next most likely storm arrival date can all be predicted by knowing about the passage of a dry front. The next time southerly winds come with partly cloudy skies but very little rain, you can be sure that the next few days will bring cooler temperatures, clearing conditions, and more northerly winds. Count three and one-half days from the arrival of this weak storm to determine the most likely time for the arrival of the next storm front.

The Jet Stream

As we have learned, it is very important to know the direction of an approaching storm. If you know the storm direction, you can predict how much of an influence the storm will have on your plans. One important clue is to know the direction of the jet stream.

First, what is the jet stream? The next time you observe high clouds such as cirrus or cirrocumulus, watch them for a few moments. See if you can notice them moving across the sky. The movement you observe is due to the jet stream. It is a fast-moving stream of air that occurs in the high level of the atmosphere. Refer to Diagram 3 and notice the wavy course the jet stream can have. This is a simplistic picture of the jet stream, but some accurate generalizations can be made.

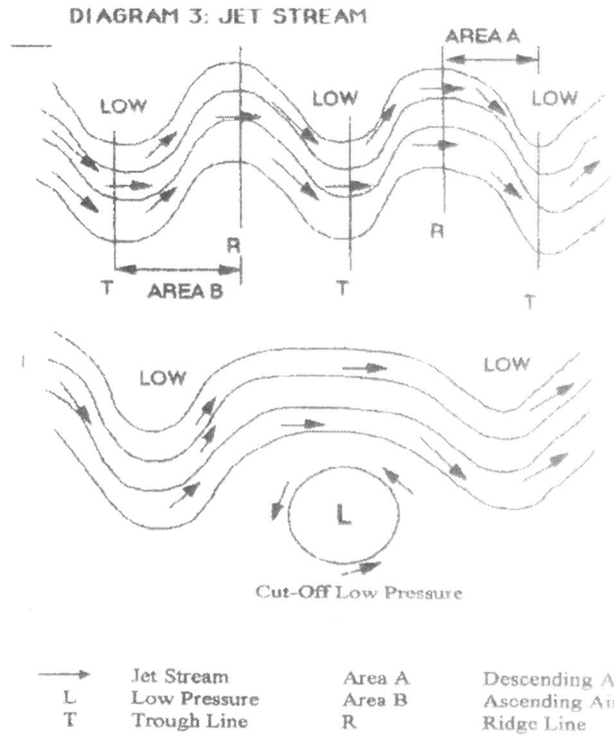

The direction, strength, and speed of a storm front is greatly influenced by these streams of fast moving air. Generally, when the jet stream is flowing from a northwesterly direction or was flowing from a southerly direction but is now westerly, expect the storm to weaken. If the jet is southwesterly or was coming from a northerly direction but is now turning westerly, the storm will increase in strength. This is due to the fact that the air in Area A of Diagram 3 generally will be descending. Air in Area B generally will be ascending. Since instability increases as air ascends, this condition will enhance the instability and strength of any approaching storm front.

Colder storms accompany a northern jet stream. Warmer storms accompany a southern jet. Expect the snow level to fall or rise respectively. The jet stream that comes out of the extreme north or south will usually force the storms that accompany them to be slow moving. Sometimes, the main flow of the jet stream will abandon a looping segment of itself (see Diagram3). Storms associated with this abandoned segment of the jet stream (called a cut-off low) have a tendency to become almost stationary because of their disassociation from the main flow of the jet stream. In this case you are usually stuck with these storms for days with unstable, cloudy, and rainy conditions.

Relative Humidity

The three layers of the atmosphere, in which clouds occur, are composed of gases that make up the air we breathe. Some of this gas is water vapor—water in the form of gas. The amount of water vapor that the air can hold is limited. The higher the temperature of the air, the greater the amount of water vapor the air can hold. The lower the temperature, the less water vapor the air can hold. This is why in extremely cold locations such as the Arctic, very little moisture falls. Storms may come, but very little water is available in the air to condense and fall to the ground. The Arctic gets about as much moisture as the deserts of the world do.

The amount of water vapor in the air is called the relative humidity and is measured as a percentage. This percentage measures the amount of water vapor that is currently in the air in relation to the maximum amount that the air can hold at its current temperature. Dew point or saturation is reached when the humidity percentage reaches one hundred percent. At dew point the air cannot hold any more water vapor. If the air was to cool when it had reached saturation, the water vapor would condense and form clouds. Clouds are collections of very small water droplets.

Changing Temperatures

What causes the air to have different temperatures? The sun warms the oceans and the land. Air will usually be warmed or cooled by the surface that it contacts. Diagram 2 shows the warm and cool air masses on different sides of the fronts. The warm or cold air masses on either side of the front get their cold or warm designations from their contrasting temperatures. If one air mass is cooler than another, then it is called a cold air mass. This has nothing to do with whether the air actually feels cold when we are outside. When an air mass is not in contact with the ocean or land but rises above those surfaces, it will generally cool. Cooling air is losing its ability to hold water vapor. As the air rises and cools because it is no longer in contact with a warming surface, its relative humidity could eventually reach one hundred percent, or saturation. At this point, the water vapor will condense and clouds will form. The time it takes to reach saturation will depend on how much water vapor was in the air mass to begin with. If the air mass had spent a lot of time in contact with a warm body or water, then the relative humidity might be high. When this humid air starts to rise and form clouds, severe weather could develop.

Stable and Unstable Air

What do you think makes air rise? In the summer, the sun heats the surface of the land. The land, in turn, heats the air in contact with it. The warmer the air becomes, the lighter it is relative to the air around it. This warmer air begins to rise just like the air over a campfire. As the warm air loses contact with the warm land, it begins to cool. If the air was somewhat humid to begin with, thunderstorms can develop. This is exactly what happens during the summer monsoons season in the western United States (see Diagram 3). The moist surface air is drawn from the Gulf of Mexico into the interior regions of the West by the clockwise winds produced by the high-pressure area found centered over southwest Colorado during that time of year. The sunshine of summer causes the moist air to warm and rise. Thunderstorms and rain are the result.

Air that is forced over mountains is made to rise and lose contact with the land. This cooling air causes the common cloudiness over the mountains when the valleys are experiencing clear conditions. This cloudiness is the reason why the mountains regions will usually receive more precipitation than non-mountainous regions. When a storm front passes over mountain regions, the mountain slopes cause the storm front and the air to rise over them. It is obvious why mountain slopes cause air to be pushed upward; it can be equally obvious why storm fronts cause the same thing once you understand a simple analogy. This analogy involves an aquarium with a removable divider cutting it in half vertically (see Diagram 5). One side is filled with water and the other side with oil. If the divider is pulled out, what would happen? Since oil is lighter than water, the water would drop down underneath the oil and the oil would rise over the top of the water. The water would go to the bottom half of the tank and the oil would rise to the upper half of the tank. The front is the area where the heavier cold air is causing the lighter warm air to rise. This rising air cools, becomes saturated, and forms clouds and precipitation.

Stable conditions occur:

1. When the air is mostly still.
2. When the air is not receiving much heat from the sun or the air is not in

contact with a warming surface.

3. When the temperature of the air increases with increased height above the ground. Usually fog will form if the air at ground level is at the saturation point.

Unstable conditions occur:

1. When winds are present that force air up and over mountains.

2. When the heat from the sun warms the land and air in contact with the land starts to warm and then rise.

3. When a front has air masses of differing temperatures in contact with each other.

DIAGRAM 4: AQUARIUM

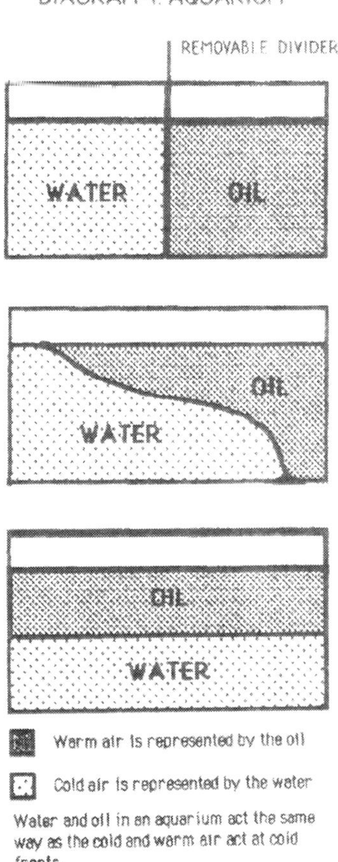

Warm air is represented by the oil. Cold air is represented by the water. Water and oil in an aquarium act the same way as the cold and warm air act at cold fronts.

Wind Is Sucked

Wind, or course, is the movement of air. What causes the air to move in the first place? As the sun shines, the air is heated but not uniformly. Air at ground level and air closer to the equator get warmed up more than air higher above the ground and air closer to the poles of the Earth.

Warmer air is lighter than colder air. This contrast in weight ultimately causes the winds to occur. Remember the movement of air caused by the warm and cold fronts in reference to the aquarium example above? The warm air mass along the ground was pushed upwards by the colder, heavier air mass. Warm equatorial air also will move northward and southward from the equator in response to the colder Arctic and Antarctic air as it moves south or northward respectively. All these displacements and movements of air cause the breezes and winds. Generally, the greater the temperature or weight difference between two air masses, the greater the wind speed.

It is common to say that the wind is blowing, but that is really not the case. It would actually be more accurate to say that the wind is sucked. Air is not pushed (blown) but rather pulled (sucked). When differences in temperature and pressure are present, air is moved by forces that tend to pull it along by creating areas of lower pressure rather than forces that push (or blow).

Getting Your Bearings

When determining the direction of the wind, knowing how to take bearings (recognizing north, south, east, and west) is going to be important. You can purchase an inexpensive compass to help you learn. When you are outdoors a compass is very helpful. Around town it may be easy to learn the directions by using familiar reference points. However, this will not work in unfamiliar surroundings.

The sun can also be used to get your bearings. The sun rises in the east and sets in the west. In the morning around nine o'clock your shadow will point approximately northwest in the United States. At noon you shadow will point north. Around, 3 o'clock in the afternoon, your shadow will point approximately northeast. When you are in an unfamiliar area or are just unsure of your bearings, establish them as soon as you can: Clouds can prevent you from getting your bearings later. By knowing only one direction you can determine all the other directions. If you face north, south will always be behind you; west will be to your left and east to your right.

Name That Wind!

Did you know that the winds have names? Winds are always named after the direction they are coming from, not the direction they are going. So, a wind that is blowing (sucked) out of the north is called a north wind. If you are in the open, the wind name can be determined by standing and turning until the wind is directly in your face. Once you can point in the direction the wind is coming from, compare that direction with north to get your bearings and to give the wind its proper name. Wind names can be determined by watching the clouds drift by a fixed point. The edge of a roof, a cliff, or any stationary object will work nicely. Watch lower-level clouds, not high clouds, if you need to know the surface wind direction. Watch high-level clouds if you need to know the jet stream direction. Flags on poles can also help determine wind names if the area around the flag is open. Buildings and other obstructions may cause the wind to swirl and the flag to give inaccurate readings. Narrow valleys and canyons can cause the direction of the winds that blow in them to be different than the regional surface wind direction.

The Coriolis Force

Air will move up or down to equalize its temperature (actually its weight) at the same level off the ground. This causes areas of high and low pressure to be formed. As discussed above, air will attempt to move from a high-pressure area to a low-pressure area.

The Coriolis force will prevent air flowing directly from high-pressure areas to low-pressure areas as expected and instead cause it to take circular routes around high and low-pressure areas. The Coriolis force is really not a force at all, but a phenomenon that occurs when air moves over the surface of a sphere such as the Earth. To experience this effect yourself, use a globe and an erasable ink marker. Spin the globe counter-clockwise. As it is spinning, touch the pen to any part of the upper (northerly) half of the globe surface, moving it in a straight line along the surface. Stop the spinning globe and observe the line you have just made. Even though you made a straight line, the line did not end up straight, it has a right-hand turn. In fact, any straight line you make while the globe is spinning will always have this same right-hand turn. This is how the Coriolis force causes the wind to circle around high pressure areas. These observations occur only in the northern hemisphere. In the south, the observations are reversed. That is why the key will only give correct forecasts in the northern hemisphere.

Types of Precipitation

There are many types of precipitation, but they are all made up mostly of water. Water will fall in the form of snow, hail, sleet, rain, and drizzle but can be categorized as: frozen water or unfrozen water. The frozen forms of precipitation (snow, hail, sleet) are caused by water droplets in the cloud forming or falling through air that is thirty-two degrees Fahrenheit or less. The unfrozen forms of precipitation (rain, drizzle) are caused by water falling through air that is greater than thirty-two degrees Fahrenheit. Sometimes precipitation will form as snow but fall through warmer air near the ground. This results in slush or rain. Water in liquid form may fall though freezing air near the ground, resulting in sleet. Hail is formed by large thunderheads and falls to the ground as balls of ice. I have observed snow falling with temperatures as high as thirty-eight degrees Fahrenheit. It formed in freezing air but fell through warmer air; the warm air near the ground was not thick enough to give the falling snow time to melt before it reached the ground. Rain can fall when surface air is less than thirty-two degrees Fahrenheit if warm air lies above. This is what happens during the passage of a warm front. If the ground temperature is cold enough or the air mass is thick enough, the rain can become sleet.

During the passage of a storm front, winds that come from the north will generally be cooler than winds that come from the south. As the storm approaches, the southern winds will bring an increase in temperature and as the storm passes, the northern winds will cause the temperature to drop. Because of this, storms that start with rain and temperatures just above freezing may bring snow and colder temperatures toward the end of the storm.

Sea Breezes

If you live along the coast, you may have noticed an interesting daily wind shift that sometimes occurs. During the night and early morning, the breezes tend to go from the land to the sea (offshore). During the day and early evening the wind has turned and has begun coming from the sea (onshore). This cycle repeats itself each day if regional conditions are stable. Knowing that cool air is heavier than warm air, what do you think causes this wind shift? Remember the aquarium example (Diagram 4) when we were studying the way air masses with different temperatures and weights impact each other? The colder and heavier air mass slides under the warmer and lighter air mass. This is exactly what happens during the day as breezes come from the sea. Since land heats up quicker than the sea during the daytime, the air over the land becomes warmer quicker than the air over the sea. Here we have the same situation that occurs along a storm front: cool air side by side with warm air. The cool, heavier air over the sea is going to slide under the warm light air over the land, causing cooling onshore breezes.

What causes the wind to shift on the coastlands as evening turns to night? It is true that the land heats up quicker than the sea during the day, but it also cools down quicker during the night. The air over the land now becomes cooler than the air over the sea. The exact reverse situation at night exists as compared to the conditions that existed during the day, therefore, offshore breezes result.

Barometers

The key will ask you periodically to take readings of the current barometric pressure. In order to do this it would be helpful to own a barometer. They are not very expensive and can usually be found at your local hardware or general store.

Most barometers will have a pointer that can be moved manually and a non-manual pointer which measures the barometric pressure. Every time you take a reading the manual pointer should be adjusted to line up with the new position of the non-manual pointer. In this way you will be able to know how the pressure is changing. It is far more important to know whether the pressure is going up or down than to know the pressure reading. Sometimes the current pressure readings can be procured by watching a local TV station. In this case you will need to record the readings to know how they are changing. As you become more familiar with the key and how to interpret the signs, you will become less dependant on the pressure readings in order to accurately predict the weather. However, a barometer will always be helpful and give you additional insight into your weather forecast. The key has been designed to be used even when you are outdoors or away from home and may not have access to the pressure reading.

In the key, the questions regarding barometric pressure are purposely placed just before reaching the outlook. This placement gives the least possible outlook alternatives if you don't know the current movement of the pressure. Some helpful hints are included in the pressure section of the key to help you answer the pressure questions without having access to the current pressure changes.

As a very general rule, the storms that bring a lot of precipitation will usually cause the pressure reading to go below 30.00". A reading below 29.60" indicates the approach of a stronger storm. Barometric pressure as low as 28.50" is more uncommon. A reading this low would indicate very stormy conditions. Storms that do not cause the reading to go below 30.00" are usually weaker. Precipitation may come but not in the amounts associated with the storms that have lower readings. High-pressure readings are those that occur above 30.00". Very high readings are above 31.00". These higher readings will usually accompany clear skies and fair conditions.

Monsoon Season and Thunderstorms

In the summertime, parts of the western half of the United States are influenced by a monsoon season (see Diagram 5). The monsoon season is caused by a high-pressure system that develops in the summer over the four corners area (the states of Arizona, Colorado, Utah, and New Mexico come together). The air in that region will blow in a clockwise rotation around the high's center. This means southeasterly winds will develop in the area between the Sierra Nevada Mountains and the Rocky Mountains, from Arizona north to Idaho. Occasionally the monsoon will be strong enough to reach west of the Cascade, Sierra Nevada, and San Bernardino Mountain chains, but that is not the norm. Sometimes the high pressure will center east of the four corners area. In this case the southerly winds will be present east of the Rocky Mountains. The north central area of the United States (from Idaho east to Iowa) will receive this monsoon moisture as the air circulates around the top of the high-pressure area. In this case, southwesterly winds forewarn of the monsoon during the summer. Moisture is brought up from the Gulf of Mexico by these winds and when it meets the summer heat of these regions, thunderclouds form, resulting in rain and thunder. Watch for cumulus clouds to develop from a clear sky and grow to extreme size. Many times altocumulus clouds or other types of clouds will forewarn of the monsoon effect. The forewarning clouds, if they occur at all, will not necessarily be in any organized predictable order, unlike the forewarning clouds of an approaching winter storm front. Sometimes several days of southerly or easterly winds will blow before enough moist air arrives to cause rain to fall. You will notice that clouds will progressively develop larger and larger each day as the monsoon develops.

DIAGRAM 5: MONSOON EFFECT IN THE WESTERN UNITED STATES

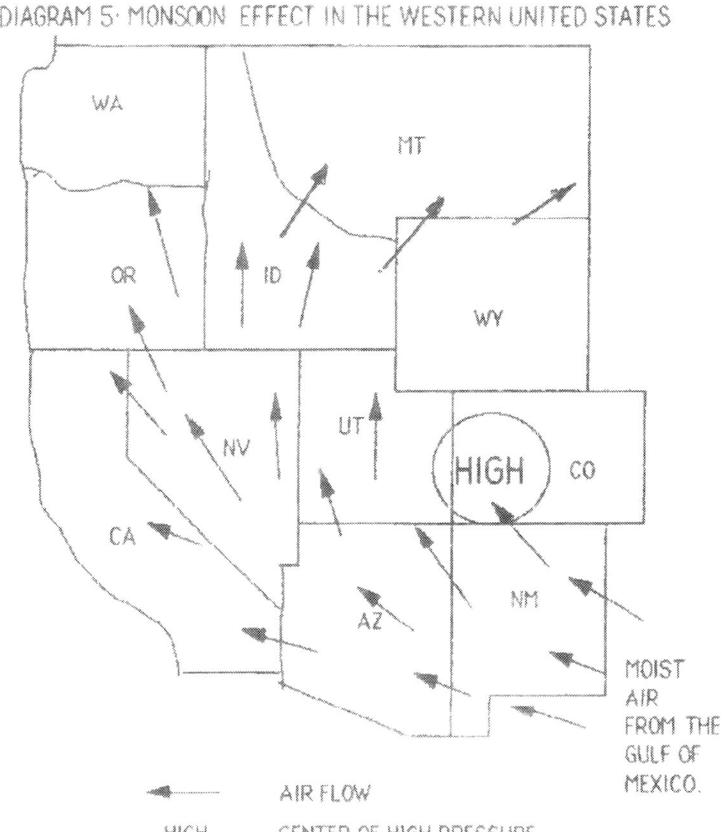

Seasonal Consideration

Fall

Fall (autumn) is the time of year when many changes occur. One needs to be especially watchful so you are not caught off guard. In the western United States, the summer weather can be uneventful. This may cause you to become less aware of the conditions around you as the fall approaches. The summer monsoon season is ending and is being replaced by a high-pressure system over the Pacific northwest or by the winter storm track. During the first part of the fall the high-pressure system is usually more dominate (and is what causes the Santa Ana winds in southern California). By late fall the winter storm track has usually become the most influential force. The jet stream flowing from west to east has begun to move farther south in the United States and with it comes the low-pressure storm systems. With the approach of the first storms of the rainy season, you will begin to observe their forewarning signs. Many times, however, the signs of the first storms will not be well developed. Various types of clouds can be missing. Instead of large and distinct formations of cloud types, you will see smaller and more fractured groupings. The barometric pressure will not fall as much as it will in the winter. Southwesterly winds will come, but the stronger storms with their south to southeasterly winds may not arrive until later in the season. The more northerly your location, the sooner the forewarning storm signs will become clearer and stronger.

Winter

Wintertime is dominated by the winter storm track. The Pacific northwest high pressure that tends to dominate in early fall has usually died away. The jet stream becomes very influential in winter. During other times of the year it travels from west to east, tracking on a more northerly course. The storms usually follow this track and do not influence the western United States as much in the summer. By late fall through winter and during the earlier parts of spring the jet steam travels far enough south to become the dominate influence affecting the weather in western United States. The eastern United States is brought storms all year by the jet stream, especially the central and northeastern areas.

Normally, a high-pressure system that sits off the coast of Oregon in the summer, moves to a more southerly location in the winter. The jet stream in the summer flows to the north of this high-pressure area. This causes most of the storms in the summer to come on shore at the West Coast of Canada. This is usually too far north to cause storminess in the western United States. In the winter, this high-pressure area normally moves south and the jet stream also moves south as it travels west to east. This brings the arrival of the storm fronts and the rainy season. If the high-pressure area stays too far to the north during the rainy season, the storm fronts end up taking a more northerly course like they do in the summer and largely miss influencing the pressure readings and also reduce the amount of precipitation received.

Spring

Spring, like fall, is a season of change. Whereas fall is the season of increasing strength of frequency of the low-pressure storm systems, spring is the time for the storms to grow weaker and less frequent. The storm forewarning signs will grow weaker as well. The primary wind directions will still be influenced for the most part by storm fronts. Winds will be southerly to east as the storm systems approach and west to northerly as the storms pass.

Toward the end of the spring the monsoon system will begin to form in the western United States. Look for the continued presence of high pressure over the four corners area of the southwest Untied States. Hotter days, southeasterly winds, and clouds of various types approaching from the east or southern horizon will forewarn of the monsoon condition developing.

Summer

In the eastern United States, the storm systems will continue to come just as they do during the winter, threatening thunderstorm activity rather than snow.

The monsoon system is the dominate force during the summer in the western United States. The further north and the further west, especially west of the Sierra Nevada Mountain line, the less influential and intense and the more intermittent the monsoon effect will be. Cloud types, pressure, and wind direction will be determined by the strength of the monsoon. The western coastlands are more influenced by ocean breezes this time of year. Normally expect onshore breezes during the daytime and off shore breezes at night. When strong offshore winds are present on the West Coast during the day, look for monsoon conditions to develop.

Planning Your Trip

If you're planning a trip, knowing the local weather conditions will help you forecast the conditions at destinations that are close to home. Generally, the weather you are experiencing will be very similar to the weather in the country around your area. Use the following rules to help you make adjustments, especially when the low-pressure storm system is the most dominate influence.

Your destination should be within five hundred miles of your home location. Greater distances will not compare well to your local conditions. Areas that are within a few hundred miles will have the same weather at the same time or will have the same weather within a few hours.

If your destination lies to the east or south (western United States only) expect to experience the conditions your home location has experienced over the last one or two days. Generally, storm system storminess will take a little longer arriving at easterly locations than southerly locations that are the same distance away from your home location.

Destinations that lie to the west or north (western United States only) will experience the conditions one or two days before your home location does. Monitor the forecast in the key for your area's forecast for one or two days. That forecast is the current conditions in areas to the west or north.

If your destination lies to the east and also lies east of a major mountain chain, expect any storm to be diminished in strength at the easterly location.

Locations with higher altitudes will experience colder temperatures and mountains will receive greater storminess.

Southerly locations will experience less storminess if the center of the storm is passing through or to the north of your home location. Look for southerly winds at the home location to confirm that the center of the storms is passing through or is passing to the north of your area.

Southerly locations will experience more storminess from storms if the center of the storm is passing south of your home location. Look for easterly winds at the home location to confirm that the center of the storm is passing to the south of your location.

If you live in the eastern United States, areas to the north or west will experience the conditions your home location has now in the next one or two days.

If you live in the eastern United States, areas to the south and west will experience the conditions one or two days before your home location does.

Conclusion

In an effort to give you a working knowledge of the weather with the least amount of work on your part, this chapter has reduced a huge subject into just a few easy-to-understand pages. Ultimately I hope this book will inspire your interest in the study of weather. You really don't need to have a degree in meteorology to amaze your friends with accurate weather predictions. Just keep using your key.

Glossary

The words in this glossary have sometimes been defined very narrowly. This limited definition focuses on its meaning associated with the weather. For a more complete definition, use any general dictionary.

Atmosphere: The gases that lie along the surface of the earth. Clouds occur in the first 60,000 feet of this layer of gasses.

Altocumulus: A cotton ball-like-cloud that is formed by vertically rising air and appears in the middle layer of the sky.

Altostratus: A grey cloud layer that is formed by air flowing up a slowly inclining plane and is in the mille layer of the sky.

Antarctic: The extreme southerly area of the earth.

Arctic: The extreme northerly area of the earth.

Ascension: To rise above the ground.

Barometer: A devise that measures atmospheric pressure.

Barometric pressure: The weight or pressure of the air.

Cascade, Sierra Nevada, and San Bernardino Mountain Chains: The mountains that form a line along the Pacific coast. These mountains separate the coastal valleys from the desert regions that lie to the east.

Cirrocumulus: A cotton ball-like cloud that is formed by vertically rising air and is in the high layer of the sky.

Cirrostratus: A grey cloud layer that is formed by air flowing up a slowly inclining plane and is in the high layer of the sky.

Cirrus: A cloud that looks like wavy hair and is in the high layer of the sky.

Clockwise rotation: Circular rotation in the direction that the hands of the clock rotate.

Cloud: Air that is saturated and cannot hold any more water vapor. The water vapor has gone from the gas state to the liquid state and has formed small water droplets. These droplets are now visible and take the form of a cloud.

Cold air mass: A body of air that is colder than another body of air. These bodies of air are usually in contact with each other.

Compass: A device that points toward the north.

Coriolis force: A phenomena that cases air that moves over the surface of the earth in the Northern Hemisphere to constantly turn to the right.

Counter-clockwise rotation: Circular rotation in the opposite direction than the hands of the clock rotate.

Cumulus: A cotton-ball-like cloud that is formed by vertically rising air and is in the low layer of the sky.

Cut-off low: A low-pressure area that has become disassociated with the jet stream.

Dissension: To fall in altitude.

Dry front: A front that brings no precipitation.

Drizzle: A light form of liquid precipitation.

Equator: The line around the center of the globe that is equal distance from the poles.

Fahrenheit scale: A scale used to measure temperature in degrees where boiling occurs at 212 degrees and freezing at thirty-two degrees.

Front: The line that separates two air masses that have different temperatures.

Forewarning clouds: These are clouds that appear as much as four days before the arrival of a front. These cloud types include cirrus, cirrocumulus, cirrostratus, alto cumulus, altostratus, cumulus, stratus, linticular, and stratocumulus.

Four-corners area: The point where the following four states come together: Arizona, Colorado, New Mexico, and Utah.

Gas: The air that we breathe that forms the atmosphere around the Earth.

Gulf of Mexico: The body of ocean water that lies between Mexico and Florida.

Hail: A form of frozen precipitation.

High layer of the sky: The highest layer of the atmosphere where clouds can occur.

High pressure: Barometric pressure readings that are above 30.00".

High-pressure system: A region where the pressure is high and the wind is flowing in a clockwise rotation.

Horizon: The line where the sky appears to meet the ground.

Intermittent precipitation: Precipitation that is showery and not falling continually.

Jet stream: A stream of air that flows in the high level of the sky that greatly influences the direction low-pressure system's travel.

Key: A way of organizing information so as to make it very practical. The reader is asked questions which will lead to other questions and then to a conclusion. The conclusion in this key is the outlook.

Lenticular: A cloud that takes the form of a stationary fog-like layer as air flows over the top of mountains or other obstructions. Many times they are saucer shaped.

Low layer of the sky: The lowest layer of the atmosphere where clouds can occur, including fog that lies along the ground.

Meteorology: The study of the weather.

Middle lay of the sky: The middle layer of the atmosphere where clouds can occur.

Monsoon season: In the summer, moist air from the Gulf of Mexico will flow into the hot interior of the western United States causing thunderstorms to form.

Mountain chains: Mountains that are grouped together in such a way so as to form along a line.

Northern Hemisphere: The area of the earth that lies north of the equator.

Off-shore breeze: A breeze that blows from land to sea.

On-shore breeze: Breezes that flow from sea to land.

Opaque: A condition where light is unable to pass through freely.

Outlook: In this book, the weather forecasts that are given at the conclusion of the key.

Precipitation: All forms of water that fall from the sky. For example: rain, snow, sleet, hail.

Rain: A form of liquid precipitation.

Relative humidity: A measurement of the amount of water being held in the air in the form of vapor.

Rings: Around the sun cirrostratus clouds will produce what appear to be rings around the sun. The clouds act to concentrate the sunlight where the ring forms to cause this visual display.

Santa Ana wind: In the fall in southern California, a condition will exist that causes these moderate to strong northerly winds to blow.

Saturate: A condition that air is in when it contains the maximum amount of water vapor possible at its current temperature.

Season: winter, summer, fall, and spring.

Sierra Nevada Mountain Line: The mountains that form a line along the Pacific coast. These mountains separate the coastal valleys from the desert regions that lie to the east.

Sleet: Partly frozen rain.

Snow: Small white crystals of frozen water formed directly from water vapor in the air.

Storm: A period of unstable weather conditions.

Storm system: A low-pressure area that has developed storm fronts.

Stratocumulus: A cloud that looks like a mixture of cotton-ball clouds and grey-layer clouds occurring in the low level of the sky.

Stratus: A grey-layer cloud that occurs in the low layer of the sky.

Sub-tropics: Areas near or bordering the tropical zone.

Surface wind: The movement of air along the ground.

Thunderhead: A large cumulus cloud that may extent to all three layers of the sky.

Tropic of Cancer: A line, parallel to and north of the equator. This line is the most northerly point at which one could observe the sun directly overhead in the summertime.

Tropics: In the Northern Hemisphere, the area that lies between the equator and the tropic of cancer.

Unstable: A condition that may lead to precipitation.

Warm air mass: A body of air that has a warm and somewhat uniform temperature. The air mass is recognized as warm because of its contrasting temperature with a nearby cooler air mass.

Warming surface: The ground or body of water that is warmer than the air that lies immediately above it. The air in contact with this warming surface warms up.

Water vapor: Water in its gaseous form.

Wind: The movement of air.

Winter storm track: The path that low-pressure storm systems will follow. This will usually be along the current path of the jet stream.

Altocumulus

Altostratus

Cirro Cirrus

Cirrocumulus

Cirrostratus

Cumulus

Linticular

Stratus

Stratocumulus

CPSIA information can be obtained at www.ICGtesting.com
Printed in the USA
LVOW11s1014290315

432470LV00001B/330/P